NO STOPPING IN THE NORTH OF BRITAIN

From Multiple Railway Stations to None

Patrick Bennett

AMBERLEY

First published 2024

Amberley Publishing
The Hill, Stroud
Gloucestershire, GL5 4EP

www.amberley-books.com

ISBN 978 1 3981 1513 2 (print)
ISBN 978 1 3981 1514 9 (ebook)

British Library Cataloguing in Publication Data.
A catalogue record for this book is available from the British Library.

Typesetting by SJmagic DESIGN SERVICES, India.
Printed in the UK.

Contents

Introduction

The pre-Grouping railway companies were fiercely competitive, the result of which was a considerable duplication of lines and stations. Each company was determined to gain its own share of passengers and goods traffic. A good example of this wasteful competition was seen in the Leen Valley where three different railway companies each built their own line winding along the valley, often within yards of their rivals, and at even quite small settlements each company built its own station. This duplication is more often to be found where there was heavy industry or mineral sources, this traffic being particularly lucrative for the railways. Duplication was not always the result of competition between companies. In some instances the same company had two or more stations in a location. At Leith, for example, the North British Company had three stations.

Unfortunately many of the towns and villages where inhabitants once had the choice of travelling from two or more stations now have none. This book recounts the history of these unfortunate places, describing how the railway and its stations came to be there in the first place and their subsequent demise. This volume looks at northern Britain.

The author and publishers would like to thank Peter Groom for permission to use the photographs taken at Staveley and at Oswestry, and to thank the Branch Line Society for permission to use the photographs of Angus McDougall. The photographs by Ben Brooksbank, Chris Allen, Phil Sangwell, Oxyman, and Lamberhurst are reproduced under the Creative Commons ShareAlike Licence 2.0. Every attempt has been made to seek permission for copyright material in this book. However, if we have inadvertently used copyright material without permission or acknowledgement we apologise and will make the necessary correction at the first opportunity.

West Yorkshire

Kilnhurst, Rawmarsh, Aldwarke, Wath, Royston, and Notton

The North Midland Railway main line opened from Derby to Masborough on 11 May 1840 and from Masborough to Leeds on 30 June. Rawmarsh station opened in May 1853 and was renamed Parkgate and Rawmarsh in December 1869. Kilnhurst opened in April 1841. The South Yorkshire Railway's line from Sheffield to Mexborough was completed in 1871. Aldwarke was opened in July 1873 and renamed Parkgate and Aldwarke on 1 November 1895. Kilnhurst opened on 1 September 1871. These two lines ran close together along the Don Valley, with the stations only yards apart. Nevertheless it was not until 1966 that the SY line lost its passenger service and Parkgate and Rawmarsh, and Kilnhust Central closed. The Midland line remains open but both Parkgate and Aldwarke, and Kilnhurst West were closed on 1 January 1968. Parkgate has since regained a station as it is now the terminus of the tram line from Sheffield.

There were three stations at Wath. The Hull and Barnsley station was the terminus of the central branch from Wrangbrook Junction. It lost its passenger service as long ago as 1929, having only been opened in 1902. The MR station was originally just Wath, followed by a number of name changes ending with its final name of Wath North. The station opened with the railway in 1841 and closed with the withdrawal of local services between Sheffield and Leeds in 1968. In 1953 it had a very sparse service, just seven trains a day. The Great Central station at Wath-on-Dearne opened in 1851. It closed on 29 June 1959 when the passenger service between Barnsley and Doncaster was withdrawn.

The Barnsley Coal Railway, later owned by the SYR, and in turn the MS&L, was opened from Stairfoot to Applehaigh in 1870 and extended to Nostell on the West Riding & Grimsby Joint Railway in 1882, at which time the station of Notton and Royston was opened. In 1895 a service of seven trains a day ran between Barnsley and Leeds, with two extra on Saturdays and just two on

Sundays. It lost its passenger service on 22 September 1930. Royston station opened with the MR main line in 1841 but was replaced by Royston and Notton a mile further south in 1901. Local stopping trains between Sheffield and Leeds via Cudworh were withdrawn on 1 January 1968.

The section of the MR main line between Swinton and Cudworth was subject to mining subsidence so in October 1968 it was decided to switch the remaining passenger traffic to the Swinton & Knottingley Railway. By 1972 the S&K was also suffering from subsidence problems so that services were switched back to the MR line, before being switched once more in the early 1980s back to the S&K.

Left: Kilnhurst Central station seen in 1976.

Below: In 1955 at Kilnhurst an ex-LNER Class 01 2-8-0 heads north with a long freight train.

At Wath Road Junction the Great Central Railway was encountered again, as depicted here. A Class 4F 0-6-0 heads south along the Midland main line with a freight train while ex-GC J11 0-6-0 No. 64363 passes underneath on the GC Wath–Mexborough line. (Ben Brooksbank)

The GC station at Wath in an early twentieth-century view.

On 24 July 1951, the Up Thames-Clyde Express passes through Royston and Notton double-headed by Caprotti Black Five 4-6-0 No. 44756 and Jubilee 4-6-0 No. 45615 *Malay States*. (Ben Brooksbank)

On 18 April 1981, when this photograph was taken just north of Royston, it can be seen that this was still a four-track railway. Class 56 No. 56074 is heading south with a train of empty MGRs probably heading for Grimethorpe Colliery. No. 56074 was one of a number of locomotives fitted with automatic slow speed control for unloading at power stations, hence the need for the orange flashing light, which can be seen atop the driving cab.

Royston Junction in 1992, now a double-track railway. The main lines between Royston Junction and Oakenshaw South Junction signal boxes were taken out of use and the goods lines renamed the main lines on 29 November 1981. The signal box was abolished in 1995.

In 1988 the MR main line was closed to all through traffic. By 1994 the main use of the line was for trains to Grimethorpe Coalite and to the glass factory at Monk Bretton. On 31 May 1994, at the site of Royston and Notton station Class 58 No. 58026 heads past with the 08.35 Middleton Towers to Monk Bretton sand train.

Travelling in the opposite direction on the same date is Class 58 No. 58012 with the ex-Grimethorpe coal empties. It was on this day that it was announced that Grimethorpe Coalite was to close. The sole use of the railway today is for the sand trains to Monk Bretton and the railway has been reduced to a single track.

Goosehill Junction was where the MR route met the L&Y line from Wakefield. The MR main line from Oakenshaw Junction to Goosehill Junction was closed on 1 June 1987, and the slow lines to Altofts Junction were taken out of use on 31 January 1988. The signal box was abolished on 2 October 1988. Its remains are seen in August the following year.

Methley

The NM main line was joined by the York & North Midland Railway at Altofts Junction simultaneously with the opening of the NM line to Leeds. The curve from Whitwood Junction to Methley Junction was completed by 29 July 1840 and for the next twenty-nine years trains from York to Leeds used this roundabout route. Methley North station opened in April 1841. Services at this time were poor, with an infrequent service to Sheffield and a slightly better one to Knottingley. The station closed on 16 September 1957.

The L&Y opened its line from Pontefract to Methley Junction on 12 September 1849. The GNR had running powers over the L&Y and MR lines and thus there were now four different companies running through Methley. This situation did not last as from 1854 onwards the companies other than the MR used alternative routes to access Leeds. The L&Y station, for some reason known as Methley Junction, although not actually at the junction, closed on 4 October 1943.

The Leeds and Methley Junction line was a joint project of the GN, L&Y, and NE. The line ran from Lofthouse Junction to Methley Joint station where it divided, one branch going to the L&Y line and the other to Methley Joint Junction on the Whitwood Junction to Methley Junction curve. The line opened for goods in August 1865 but not until July 1869 for passengers. Methley Joint station, also known as Methley South, closed on 7 March 1960, leaving Methley without an active station.

Methley Joint station. The line to the left was the L&Y line to Pontefract. The line on the right led to Whitwood Junction and the NER line to Castleton.

Methley North, the MR station. A Fowler 4MT heads south.

Methley Junction, the L&Y station, seen here, was just to the south of Methley Junction.

Ecclesfield, Heckmondwike, Cleckheaton, Liversedge, Northorpe, Gildersome

There was duplication elsewhere in West Yorkshire. The GC and MR followed each other closely along the Blackburn Valley. Each had a station at Ecclesfield. The MR station, Ecclesfield West, closed on 6 June 1967. Ecclesfield East, the GC station, closed on 7 December 1953.

Duplication was also to be found in the Spen Valley, with this time the L&Y and LNW competing for the same traffic. First to open on 12 July 1848 was the 7-mile L&Y Spen Valley line from Mirfield to Low Moor. There were stations at Heckmondwike, Liversedge and Cleckheaton. Northorpe opened in 1891. Heckmondwike became a junction with the opening of the branch to Thornhill in 1869. From 1883 the GNR had running powers over the line, the trains often hauled by L&Y engines. Passenger services were withdrawn in 1965 but freight traffic continued into the 1980s. The LNW also had running powers over the Spen Valley line, initially only for trains running express, but it did not have running powers for freight trains. In the light of this and the congestion and delays on the Spen Valley line, in 1900 it opened the Heaton Lodge & Wortley Railway to provide an alternative route between Huddersfield and Leeds. The LNW also had stations at Northorpe, Heckmondwike, Liversedge and Cleckheaton, all of which closed to passengers on 5 October 1953.

Further north, the Leeds New Line met the GN Leeds, Bradford & Halifax Junction Railway at Gildersome. The LNW station closed as a wartime economy measure on 1 August 1917 and reopened on 5 May 1919, only for it to close completely on 11 July 1921. The GN station was opened on 20 August 1856. Gildersome West, as it had become known, closed to passengers on 11 June 1955 and the line itself closed progressively from 1966.

Cleckheaton Yard is seen in the 1960s, still busy with freight. Notice the brake tender behind the Class 37, used to give additional brake force with trains of unfitted wagons. Also of note is the coal merchant filling his hundredweight bags, once a common sight at every station.

Cleckheaton Spen, the LNW station.

The LNW cutting through Heckmondwike with some of its seven bridges. This is now part of a green walk.

Heckmondwike L&Y station seen in a nineteenth-century view.

A short westbound freight arrives at the disused Gildersome West station behind a Class 4 2-6-0.

Ingrow

The Keighley & Worth Valley Railway was authorised by an Act of 30 June 1862. It was constructed as a single line and opened in April 1867. It became part of the MR in 1881. In 1895 the line enjoyed an unbalanced service of eleven trains to Oxenhope and fourteen in the reverse direction. There were five trains on Sundays. By 1953 there was still a good service, with ten trains on weekdays and fourteen on Saturdays. This latter was no doubt due to the popularity of Haworth as a destination. Diesel multiple units took over the service in 1960 but they were unable to save the line, which closed altogether in 1962. As is now well known the railway reopened as a preserved line in 1968. The station at Ingrow was demolished. The present structure is Foulridge station, taken apart and reconstructed at Ingrow.

The GNR had ambitions in the West Riding and this included building a line to Keighley. By means of the Halifax Thornton and Keighley Railway Act of 5 August 1873 the GNR was able to purchase the Halifax & Ovenden Railway and construct lines to Bradford and Keighley. This involved considerable difficulties and it was not until 1884 that the railway reached Keighley, where a new station was shared with the MR. Passenger services finished on 23 May 1955 and goods services beyond Ingrow to Cullingworth ended in 1956. Goods services to Ingrow continued until 1965.

Ingrow West. (Chris Allen)

Ingrow East seen in the 1950s.

Hemsworth, Crofton

Promoted by the South Yorkshire Railway and River Don Navigation, the West Riding, Hull & Grimsby Railway was authorised in 1862. It extended from Wakefield Westgate to Stainforth Junction, and from Adwick Junction to Doncaster. The line to Doncaster opened on 1 February 1866, and the section from Adwick to Stainforth on 1 November the same year. Also in that year the railway was vested jointly in the GNR and MSL as the West Riding & Grimsby Joint Railway. Hemsworth was one of the original stations. For a village station it had quite a reasonable service in 1895. There were eight trains in each direction running between Doncaster and Leeds, Bradford or Halifax. In addition one of the Up London trains passing through stopped to pick up passengers, and two of the Down London trains stopped to set down 'on notice being given to the guard at Doncaster'. There were three trains on Sundays. By 1960 there were nine Down trains and eleven Up, some of these being operated by diesel units. One of the Up services ran through to Cleethorpes. There were no Sunday trains. The station closed on 6 November 1967.

The other Hemsworth station was Hemsworth and South Kirkby and was on the Hull & Barnsley Railway line to Stairfoot. The station was located to the south of Hemsworth and a considerable distance north of South Kirkby. In 1895 there were six trains daily in each direction, with just one on Sundays. The passenger service between Howden and Cudworth was withdrawn on

1 January 1932, and the line as far as Wrangbrook closed completely in 1967. The station buildings are seen sometime after closure.

Further north on the WRG is Hare Park and Crofton station, which did not open until November 1885. It closed to passengers on 4 February 1952. Hare Park was just a large country house to the south of the station. The Wakefield, Pontefract & Goole Railway opened its line on 1 April 1848, by which time it was part of the Manchester & Leeds Railway, later L&Y. Crofton Station opened on 1 November1853. It was further away from the village than Hare Park and it was no surprise that it closed as early as 30 November 1931.

Hemsworth station.

Ryhope, Drax, Pateley Bridge

Ryhope

Looking at a map of the railways of north-eastern England it is clear that there was very little duplication in the provision of stations and the reason is clear. The whole area from the Tweed to the Humber was dominated by one railway company, the North Eastern Railway.

On 8 February 1853, the Marquess of Londonderry started the construction of a railway from Seaham to Sunderland, the main purpose of which was the transport of coal from his mines. This 7-mile-long line was almost entirely on the Marquess's own land. The carrying of coal to the staithes at Sunderland started on 3 August 1854. Passengers were carried from 2 July 1855. There was an intermediate station at Ryhope. In 1899 the railway possessed thirteen locomotives, some built in its own workshops, and 2,862 goods and mineral wagons. In that year it carried over half a million passengers and 869,000 tons of coal. The passenger service consisted of twelve return trips on weekdays and six on Sundays. In 1900 the railway was purchased by the NER for £400,000. A condition of the purchase was that the Marquess could stop any train, other than expresses, at his private station at Seaham Dene. This right continued to be exercised until 1925. The line remains open today, having later been extended south to Hartlepool by the NER, but Ryhope station closed to passengers in 1960, and to goods in 1964.

The Durham & Sunderland Railway obtained its Act on 13 August 1834. It opened for mineral traffic in July 1836 and for passenger traffic between Ryhope and Sunderland on 19 October the same year. The railway was extended to Shincliffe in 1839. It was worked entirely by stationary engines until 1855, after which loco haulage was used exclusively. The railway was purchased by the York & Newcastle Railway on 27 July 1846, the Y&N later becoming part of the NER. Despite being worked by locomotives the steep gradients had not been eliminated, in particular the 2½ miles at 1 in 60 and

1 in 44 descending to Ryhope. On 19 August 1889, there was a serious accident when a Liverpool to Newcastle via Sunderland express derailed on the curve through Ryhope station. It was deemed that excessive speed was responsible for the derailment and the injuries to 104 passengers. In 1931 the passenger service was cut back to Pittington, ceasing altogether on 5 January 1953. The line continued to carry freight, particularly coal, latterly from Murton Colliery. It closed altogether in 1991.

Left: Ryhope East station. Notice the station master's wonderful vegetable garden. Just visible on the left is the end of the coal drops.

Below: Ryhope Grange Junction in 1993 when the line from Murton was still active. The branch has now been lifted.

On 1 January 1993 Class 60 No. 60028 *John Flamsteed* heads through Ryhope Grange Junction with a train of oil tanks.

Drax

There is a rule of thumb about railway closures, which is that the last to be built were often the first to close. Later lines often served smaller communities or were to some extent duplicate lines. Both cases are true of the Hull & Barnsley Railway. The impetus for the building of the H&B lay in the resentment of the predominant position of the NER for the traffic to the port of Hull. At a meeting at the Station Hotel in Hull a plan was put forward for both a railway and a new dock. Thus it was that the Hull, Barnsley & West Riding Junction Railway and Dock Company was formed. Its main activity was to be the transport of coal from the South Yorkshire coalfields to a new dock at Hull. The Act received royal assent on 26 August 1880. Construction took five years and was even halted at one point due to lack of funds. The line had cost more than double the original estimate to build. In 1895 Drax had a similar service to Hemsworth except that passengers could avail themselves of two additional services by signalling to the driver. In 1922 the H&B was absorbed by the NER. As mentioned above, passenger services were withdrawn between South Howden and Cudworth in 1932. All remaining passenger services were withdrawn on 1 August 1955, leaving the H&B as a freight-only line. In 1959 came closure to all traffic between Little Weighton and Wrangbrook Junction. The rest of the system closed progressively between 1959 and 1968. Today, there just remain a short section of line in Hull and the branch serving Drax Power Station.

A notable absence from the NER tiled wall map of 1903, seen at many NER stations, is the Selby–Goole line. Perhaps by this time the NER was confident that its network was complete. In fact it was only when a rival scheme was proposed that the company finally got round to applying for its Act. Interestingly, the route of the line was almost exactly the same as one proposed by George Hudson some sixty-five years previously. It left the Leeds to Selby line at Thorpe Gates, crossed the ECML by a bridge to reach Brayton East Junction where there was a connecting spur to the main line. The first station was at Barlow. At Camblesforth there was just a siding accessed by a ground frame. The next station was Drax Hales, giving the little village of Drax its second station.

Work started in 1907 and was completed by 1910, when the first freight trains ran. Passenger services started on 1 May 1912. The initial passenger service consisted of three Selby–Goole returns per day, with an additional train on Selby market day. Services were operated by what the NER called 'Autocars'. These were push-pull trains consisting of a locomotive and carriages. By 1916 there was a second additional train on Selby market days.

Before too long the line was singled, probably shortly after the creation of the LNER. This left the only signal box and passing place at Barlow. After the 'Autocars' came the Sentinel steam cars, introduced by the LNER in the late 1920s. In the later years the habitual motive power was in the shape of a 'Railmotor', a G5 0-4-4T plus carriages, operating in push-pull mode. In 1947 there were four trains from Selby to Goole and five in the reverse direction. In 1957 steam gave way to Class 104 diesel multiple units. In that year there were five returns on weekdays plus an additional service on Wednesdays. It is interesting to note that the journey time of twenty-four minutes was exactly as it had been in 1916! The end came in 1964 with the withdrawal of all services.

The station buildings at Drax Abbey might seem rather too opulent for a railway that was mainly built to carry coal. On 3 April 1959, WD 2-8-0 No. 90571 carries out shunting operations. (N. Stead)

The rather basic set-up at Drax Hales. (N. Stead)

Pateley Bridge

The NER opened its Nidd Valley Branch on 1 May 1862. It ran from Ripley Junction, on the Harrogate–Ripley line, to Pateley Bridge, a distance of 11½ miles. There were three intermediate stations, to which two were added later. At Pateley Bridge there was a goods yard, engine shed and turntable. There was also a connecting line to the Nidd Valley Light Railway. The initial service was of four trains daily in each direction, running between Harrogate and Pateley Bridge. By 1905 there was an additional service and by the 1920s two more. By 1948 there were once again just four trains daily, and in 1951 passenger trains ceased altogether. Freight continued until 1964.

Power and Traction Limited applied for a Light Railway Order under the Act of 1896 to build a 7¼-mile, 3-feet 6-inch-gauge line from Pateley Bridge to Lofthouse. Finance was slow in forthcoming and in an Order of 1 March 1904 the undertaking was transferred to the Bradford Corporation. The Corporation required the railway for the construction of two reservoirs further up the valley. Work began in July 1904 to construct the railway to standard gauge. It opened to both goods and passengers on 11 September 1907. Two locomotives were purchased from the Metropolitan Railway. These were Beyer-Peacock 4-4-0Ts Nos 23 and 34. They were renumbered and named No. 1 *Holdsworth* and No. 2 *Milner*. Various other locomotives were bought later. A 4-mile extension from Lofthouse to Angram was built by the Corporation to enable the construction of the Angram and Scar House reservoirs, which were completed by 1931. In the meantime the railway had closed to passengers in 1929, a victim of bus competition. The railway closed altogether in 1936 and all the materials were sold off by auction.

Pateley Bridge NER station in 2021. Apart from the loss of the canopy, it is more or less in original condition.

At Pateley Bridge NVLR Avonside Engine Company 0-6-0ST *Blythe,* built in 1922, waits to leave just one year before passenger trains ceased.

The Leen Valley

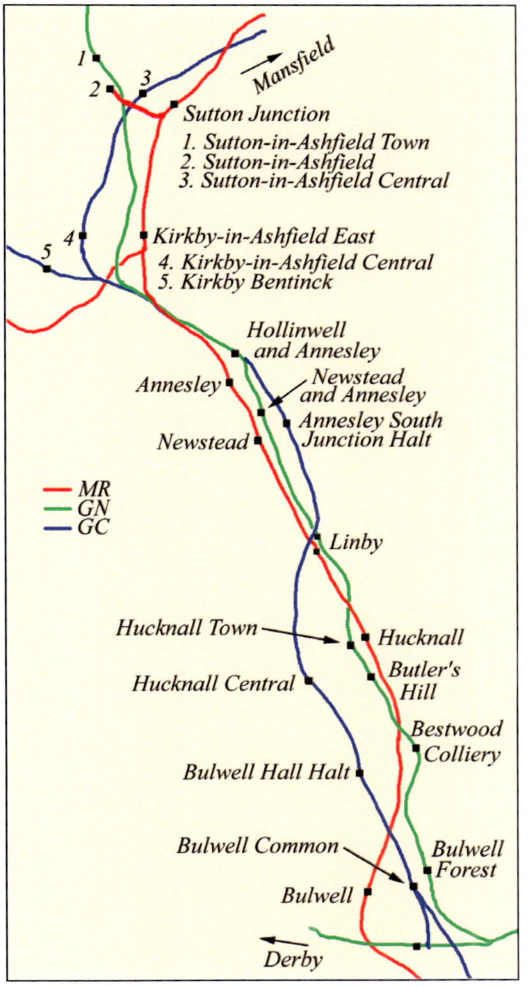

The Leen Valley.

Probably the greatest concentration of duplicated lines and stations anywhere is to be found on the borders of Derbyshire and Nottinghamshire, brought about by the intense competition for the movement of coal, a lucrative traffic for the railways. The railways concerned, the MR, GNR, and M&SLR, had less interest in the transport of passengers but nevertheless stations were opened on every line. Nowhere was this wasteful competition more evident than in the Leen Valley.

The Midland Railway opened its line as far as Kirkby-in-Ashfield on 2 October 1848. Its main purpose was the transport of coal from the collieries along the valley but there were passenger stations at Bulwell, Hucknall, Linby, Annesley, and Kirkby-in-Ashfield. The line was extended to Mansfield in 1850.

The Great Northern Railway was keen to get involved in the lucrative coal traffic and an Act was obtained in August 1880 for a line from a junction on the Derbyshire Extension Line (Leen Valley Junction) to Newstead. There were stations at Bulwell Forest, Bestwood, Butler's Hill, Hucknall Town, Linby, and Newstead. These stations opened on 2 October 1882 but coal trains had been running from Bestwood Colliery since July 1881 and the line was opened throughout for coal traffic on 27 October 1881. In 1892/93 the GNR obtained Acts for the construction of the Leen Valley Extension Line. This line would necessitate the driving of a tunnel through the Robin Hood Hills. The Manchester, Sheffield & Lincoln Railway was at that time building their extension from Woodhouse Junction to Annesley and it was decided that the 198-yard tunnel would be shared. Originally known as Shoulder of Mutton Tunnel, it was later renamed Kirkby Tunnel. The Leen Valley Extension finally opened throughout to Langwith Junction in May 1901.

The MSLR obtained an Act on 26 July 1889 to build a branch from Woodhouse Junction to an end-on junction with the GNR at Annesley. The line opened on 24 October 1892. The branch was built ostensibly to gain access to the lucrative Leen Valley coal traffic – by the end of the nineteenth century there were collieries operating at Bulwell, Bestwood, Hucknall (Nos 1 and 2), Linby, Newstead, Annesley, and Bentinck. In fact it was the first move in the building of the MSLR's London Extension. Permission to build this line was obtained by an Act of 28 March 1893. Construction work started in 1895 with the line opening for coal traffic on 25 July 1898 and to passenger traffic on 15 March 1899. On that date the stations at Bulwell Common and Hucknall came into use. Kirkby and Pinxton (later Kirkby Bentinck) had opened a few days earlier. Hollinwell and Annesley opened on 1 November 1901, and Bulwell Hall Halt on 24 April 1909. In 1897 the MSLR changed its name to the Great Central Railway.

The last development in the area was the Mansfield Railway. Nominally an independent company but backed by the GCR and operated by that company from the outset, it received its Act in 1910 and was opened throughout on 4 September 1916. The 11-mile line ran from Kirkby-in-Ashfield to Clipstone. Three new stations were opened, at Kirkby-in-Ashfield, Sutton-in Ashfield, and Mansfield, and passenger services started on 2 April 1917.

With three railways providing passenger services there was clearly an overprovision of stations. In 1861 Linby had a population of just 257 so it is unsurprising that the GNR station succumbed on 1 January 1916, with Bulwell Forest following in 1929. Local passenger services on the GNR line ceased altogether on 14 September 1931 but the line remained open for freight.

The GCR built a large freight yard at Annesley together with a locomotive depot. The six-road depot opened in 1898. It housed predominantly freight locomotives. In 1954, of the allocation of seventy-five engines, fifty-three were Robinson 2-8-0s. These were largely replaced by Standard 9F 2-10-0s, which became renowned for their work on the fast freight 'Windcutters'. The Annesley 'Dido' was a train run for the staff at Annesley, where nearly 1,000 were employed. It ran from Bulwell to either Newstead GNR or Hollinwell and Annesley. Annesley South Platform was opened as a staff halt in 1923. It closed in 1962 when the 'Dido' stopped running. In 1965 the yard closed, followed by the depot on 3 January 1966. The passenger service on the GC line was withdrawn on 4 March 1963 and the line closed entirely on 5 September 1965.

Passenger trains on the MR line ended not long after those on the GCR line, being withdrawn on 12 October 1964. Freight continued but the line was cut back to Annesley in 1968. There had been huge investment in the Leen Valley Mines in the 1950s and coal was thought to have a bright future. Bentinck Colliery started producing in 1896. In the 1980s most of the production went to Radcliffe-on-Soar Power Station. In 1988 it was merged with Annesley. Annesley-Bentinck closed in 1994 but was reopened in private hands the same year. Production ceased altogether in January 2000.

A new colliery was opened at Calverton in 1952, which necessitated the construction of a 7-mile branch from Bestwood Junction. The other collieries closed one by one: Bestwood in 1971, Hucknall in 1986, Newstead in 1987, and Linby in 1988. Calverton closed in 1993 but was reopened a year later in private hands. It continued in production until 1999.

The line from Sutton Junction to Sutton-in-Ashfield (MR) closed in 1949. The GN station, Sutton-in-Ashfield Town, had closed with the withdrawal of the passenger service in 1931. Sutton–in-Ashfield Central closed on 2 January 1956. On 20 February, it was decided to reinstate a service of half a dozen trains a day between Sutton-in-Ashfield Town and Nottingham. However, the continuation of this service was conditional on sufficient patronage. Clearly this was not forthcoming as the service was once more withdrawn on 15 September the same year.

Although the southern part of the Leen Valley line was closed, the section northwards from Kirkby remained in use for coal traffic. Trains travelled north via Worksop or south via Pye Bridge. This is the view north at Sutton Junction in 1991. The platforms were behind the photographer. The new Sutton Parkway station is about half a mile to the south.

Sutton Town station is seen in this early photograph. Huthwaite was a village about a mile away to the west.

At Kirkby-in-Ashfield East on 13 July 1963 ex-MR 4F 0-6-0 No. 44531 is seen with a Down freight. (Ben Brooksbank)

Above: Not long before closure, on 28 April 1999 Class 58 No. 58032 *Thoresby Colliery* departs from Bentinck Colliery with a loaded MGR.

Right: On 28 March 1956, a freight train heads south on the ex-GN line at Linby behind B1 4-6-0 No. 61056. In the background a train passes on the ex-GC line. (Ben Brooksbank)

Linby ex-GN station in 1990. The line closed completely in 1968.

Following closure of Linby Colliery trains went no further north than Bestwood Park Junction. This is a view north at Linby in 1990. The lines on the right are coming from Linby Colliery and the signal box is Linby. The station buildings were on the right.

The 'Dido' is seen at Hucknall Central in the charge of ex-GC Class F4 2-4-2T No. 7107.

At the GN Hucknall station G1 0-4-4T No. 630 arrives to be greeted by a parade of the Boys' Brigade, no doubt waiting for some dignitary.

Bestwood Park Junction, seen from the south in 1990. The line to the right leads to Calverton Colliery. Calverton closed in 1993.

A similar view taken on 28 April 1999. It is as well that the MR line remained in use for colliery traffic for it was to form the basis of a new passenger service between Nottingham and Worksop via Mansfield. The service started between Nottingham and Newstead in 1993, was extended as far as Mansfield in 1995, and to Worksop in 1998. Another development has been the NET tram line, which runs as far as Hucknall. This opened in 2004. A Class 153 plus a Class 150 pass with a service for Worksop.

Bulwell Common station seen after closure. Notice the enormous number of sidings. (RCTS)

Looking south at the MR Bullwell station, renamed Bullwell Market in 1952. The photograph dates from 21 August 1963. The station closed to passengers on 12 October the following year. (Ben Brooksbank)

In May 1946 D2 4-4-0 No. 3047 stands at the GN Basford and Bullwell station (later Basford North) with a service for Derby.

Seventeen years later on 21 August 1963 the 17.18 Nottingham to Derby Friargate is in charge of Ivatt 4MT 2-6-0 No. 43059. The station closed the following year. (Ben Brooksbank)

A rather desolate scene at Basford North in October 1968. (BLS)

Basford MR station was renamed Basford Vernon in 1952. It is seen here three years after closure. (Ben Brooksbank)

The GC station of New Basford closed on 7 September 1964.

The Erewash Valley

The MR line from Long Eaton Junction to Codnor Park opened on 6 September 1847. The line from Codnor Park to Mansfield via Pinxton was built largely on the formation of the Mansfield & Pinxton Railway, a horse-drawn tramway opened in 1819 and bought by the MR in 1848. Pye Bridge to Clay Cross was opened on 1 May 1862. Branches were opened to Shipley in 1870, to Ilkeston Town in 1879, and to Heanor in 1895. These branches did not fare well. First to close in 1926 was the MR station at Heanor, which was outlived by the GNR station for another thirteen years, the latter closing in December 1939. However, for the final eleven years this was an unadvertised service for workers travelling from Nottingham to work in a hosiery factory at Heanor. Ilkeston Town closed on 16 June 1947. There was a curious situation on the MR line to Heanor. A branch led off at Heanor Goods Junction to a station adjacent to the Langley Mill station on the main line. This was also called Langley Mill. It opened in 1895, closed temporarily in 1917, and closed permanently in 1926.

The GNR had long been frustrated in their attempts to gain access to the lucrative coal traffic emanating from the Erewash Valley. The company brought forward a plan for a line from Colwick curving north round Nottingham as far as Awsworth Junction, to the west of Kimberley, where a branch would head up the valley to Codnor Park and Pinxton. Authorisation was received on 25 July 1872 and the line to Pinxton was opened on 23 August 1875, initially for coal traffic. Passenger trains started running on 1 August the following year. A branch from Ilkeston to Shipley, initially for coal traffic, was opened in June 1886. This was extended to Heanor in July 1891. The station was known unofficially as Heanor Gate to distinguish it from the MR station. In 1950 it was renamed Heanor South. Passenger services were withdrawn in 1939. Other stations were at Newthorpe, Eastwood and Langley Mill, Codnor Park, Pye Hill and Somercotes, and Pinxton. Passenger services on the Pinxton Branch were withdrawn on 7 January 1963. Ilkeston North, as it had become, closed on 7 September 1964. When the GNR station of Eastwood and Langley Mill opened in 1876, the MR renamed their main-line station Langley Mill and Eastwood.

Not a great situation for the people of the Erewash Valley, but all was not lost. Langley Mill station reopened on 12 May 1986, and Ilkeston station on the site of the old Ilkeston Junction on 2 April 2017.

Pinxton South station seen in 1956. (BLS)

To the south-west of Pinxton GN station was Pinxton loco depot, seen on the right in this photograph. Stirling GN Class E2 2-4-0 No. 704 takes water.

At Pinxton on 20 July 1991 Class 58 No. 58030 at head of 7Z83 sets off after a signal check. The colour light signal is Sleights East distant. The MR station was on the other side of the crossing.

The GN station at Codnor Park. 'For Ironville and Jacksdale' was added on 22 May 1901.

At Codnor Park and Ironville, the MR station, on 19 September 1959, Ivatt 2MT 2-6-2T No. 41320 pauses with the RCTS Notts and Derbyshire railtour. (BLS)

At Langley Mill and Eastwood on 13 April 1960 an Up train of empties is seen behind LMS 4F 0-6-0 No. 44321. (Ben Brooksbank)

In 1910 a Stirling GN 0-4-4BTP stands at Heanor station, known colloquially as Heanor Gate and under BR Heanor South.

Much later the RCTS railtour visits the closed station. (BLS)

Ivatt Class D2 4-4-0 No. 1363 rolls into Ilkeston with a service for Derby.

Ilkeston Town, the MR station. Notice the complete absence of buffers.

North Derbyshire

The Erewash Valley line from Pye Bridge to Clay Cross was opened by the MR on 1 May 1862. The railway from Westhouses to Teversall and Pleasley opened on 1 May 1886. The Doe Lea line from Seymour Junction to Pleasley opened over a period of twenty-seven years. First came the section between Seymour Junction and Bolsover Woodhouse in May 1866, followed by the extension to Glapwell twenty years later. Finally the line to Pleasley was opened on 1 September 1890. There were several coal mines along its length, including Markham, the largest coal mine in North Derbyshire.

Seymour Junction to Elmton and Cresswell opened in 1888. Clowne and Barlborough was the one intermediate station. The station went through various name changes durng its lifetime before it finally closed to passengers on 5 July 1954. In its last year it had just one train daily, mainly for the benefit of schoolchildren, running between Chesterfield and Elmton and Cresswell, extended to Shirebrook West in term time. On summer Saturdays there was a through train to Blackpool and these trains continued until 1962. Although the rails have been lifted the trackbed between Seymour Junction and Oxcroft has been protected for possible future use.

It can be seen that in North Derbyshire the Midland Railway with its network of lines had a monopoly on traffic, both freight and passenger, but towards the end of the nineteenth century things began to change. In the 1880s the MS&L aspired to reach Chesterfield and to access the North Midland Coalfield. Its first attempt was thwarted by Parliament but if finally obtained its Act in 1889. This Act provided for a line from Woodhouse Junction to Annesley, where it would connect with the GNR Leen Valley line, and a branch from Staveley to Chesterfield. A further Act provided for a line from Chesterfield southwards regaining the Annesley line at Heath. This became known as the Chesterfield Loop. These lines were collectively known as the MS&L/GC Derbyshire Lines. Beighton to Staveley opened on 1 June 1892 and the branch to Chesterfield a few days later. The line through to Anneseley was opened on 2 January 1893, and the loop to Heath on 3 July the same year. These new lines not only gave access to the Nottinghamshire collieries but also provided a route to London via the GNR system.

More competition arrived with the setting up of the Lancashire, Derbyshire & East Coast Railway. The LD&ECR originally envisaged a network of some 170 miles, stretching from Warrington in the west to Sutton-on-Sea in the east. The main purpose of the railway would be the movement of coal west to the

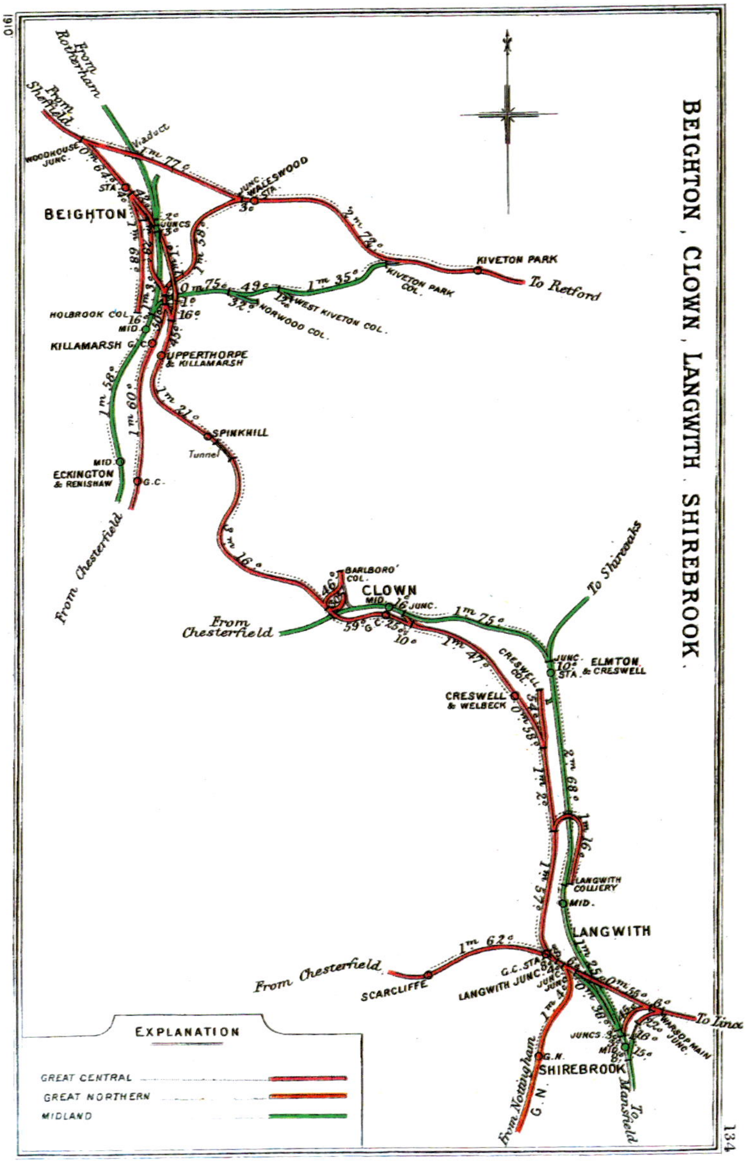

This clearing house map shows the multiple stations of Killamarsh, Eckington, Clowne, Creswell, and Shirebrook.

industrial north-west and east for export at a new harbour to be built at Sutton. The company received its Act on 5 August 1891 authorising the construction of a line from Chesterfield to Pyewipe Junction, Lincoln. Barlborough Colliery to Pyewipe Junction opened in November 1896 and the remainder of the line in March 1897. The terminus in Chesterfield was at Chesterfield Market Place. Although built primarily as line for the transport of coal, a passenger service was also provided.

The final penetration into the area was by the GNR's Leen Valley Extension Line, opened on 1 November 1901. The line was built for coal traffic and passenger traffic was never important. There were stations at Pleasley, Skegby, and a branch to Teversall.

Right: Although Killamarsh West closed to passengers in 1954 the station still saw the passage of holiday trains like this 09.25 Filey Holiday Camp to Kings Norton seen passing on 17 July 1963 behind Jubilee 4-6-0 No. 45676 *Codrington*. (Ben Brooksbank)

Below: At the GC station of Killamarsh Central on 23 August 1963 O1 2-8-0 No. 63786 heads south with a rake of mineral empties. (Ben Brooksbank)

The third Killamarsh station was the former LD&EC station of Upperthorpe and Killamarsh. The station master poses with his family.

The GC and MR each had a station named Eckington and Renishaw. The GC station was renamed Renishaw in 1950. It closed on 4 March 1963.

The MR Eckington and Renishaw closed in 1951.

The GC and MR stations of Staveley Town were next to each other. Just south of the GC station, which later became Staveley Central, was the locomotive depot of 41H Staveley, where we see D11 Director 4-4-0 No. 62663 *Prince Albert* on 15 May 1959. (Peter Groom)

The MR Staveley Town station in 1956. (Phil Sangwell)

The closed Clowne MR station in 1990. The line remained open for freight.

A wintry day at Creswell, later renamed Elmton and Creswell. The station closed on 12 October 1964. The GC station of Creswell and Welbeck closed in 1939.

At Elmton and Creswell Junction on 13 March 1992 Class 56 No. 56003 heads south with a loaded MGR. The line behind the signal box led to Clowne and Staveley. It has since been lifted. Elmton and Creswell station was to the south of this point.

There were three stations at Shirebrook. Shirebrook West was the MR station. On 17 June 1957, 2P 4-4-0 passes through the station with an inspection saloon. (Ben Brooksbank)

Shirebrook North, the GC station, until 1924 was known as Langwith Junction. On 26 May 1966, B1 4-6-0 No. 61302 stands with the RCTS Eight Counties Railtour. Shirebrook South was the GN station. (Ben Brooksbank)

At Tibshelf Sidings on 13 June 1957 ex-MR 3F No. 43235 heads through with a Down empties. Known as Tibshelf and Newton, the MR station closed in 1930. The GC station of Tibshelf Town closed in 1963.

Teversall GN was at the end of a branch from Skegby Junction on the Leen Valley Extension Line. It did not appear in the public timetable and was for the benefit of miners working at Silverhill and Teversall collieries. It is seen on the occasion of an SLS tour. The MR station closed in 1930 but remained open for excursion traffic until 1963.

The North West

Garstang

The Lancaster & Preston Junction Railway opened on 25 June 1840. There were eight stations, including Garstang. The L&PJ was in trouble from the outset, suffering from competition with the Lancaster Union Canal, and it was the canal company which leased the railway in 1842. In 1848 the Lancaster & Carlisle Railway took over both the canal and the L&PJ. In 1859 the L&C became part of the LNW. It is now part of the West Coast Main Line. In 1881 Garstang was renamed Garstang and Catterall. In 1895 there was a weekday service of six trains each way with an additional service on Saturdays. There were two trains on Sundays. By 1953 this had reduced to four trains on weekdays and two on Sundays. The station closed to passenger traffic on 3 February 1969.

The Act for the Garstang & Knott End Railway was passed on 30 June 1864 but it was to be another five-and-a-half years before the first section of line, between Garstang and Pilling, opened on 5 December 1870. The railway was soon in financial trouble and closed completely in 1872. It managed to reopen on 17 April 1875 but it was only in 1908 that the line was finally extended to Knott End. At the Grouping the railway became part of the LMS and on 29 March 1930 passenger services were withdrawn. In the final years there were five or six services in each direction on weekdays plus two more between Garstang Town and the main-line station. These services were mostly in the hands of LNW rail motor No. 10698. These machines had been introduced in 1905. They were powered by a two-cylinder engine driving the leading bogie. The engine and bogie could be removed as one unit when maintenance was needed. There was an important traffic from the saltworks at Preesall, usually handled by either ex-L&Y or ex-LNW 0-6-2Ts. After closure to passengers, freight services continued but the railway beyond Pilling was closed in 1950. Freight was later handled by ex-LNW 0-8-0s and Standard Class 2 2-6-0s. Final closure came in

An early photograph of Garstang Town. A Hudswell Clarke 0-6-0ST approaches the station.

Ex-LNW rail motor No. 10698 is seen at Knott End.

A busy moment at Garstang and Catterall, *c.* 1905.

1965, by which time only the section between Garstang Town and the main-line station was still active.

Winsford, Over

The 6½-mile branch from Cuddington on the CLC line to Chester opened in 1870, closed to passenger traffic from 1873 to 1886, then reopened to passengers for a further two years. Following an accident in 1888 the railway was required to spend a considerable sum to return the line to the required standard. Being disinclined to do so the company chose to close the line to passengers instead. The Winsford Local Board appealed to the Railway and Canal Commission, which ordered that the works be carried out and the line reopened to passengers, which it did in 1892. There was a service of six trains a day, also serving the intermediate station of Whitegate. In 1930 passenger trains were once again withdrawn, while freight continued until 1958.

The 1-mile branch to Over and Wharton was opened by the LNW in 1882. Its primary purpose was the transport of salt from the main works located to the north of Winsford. Passenger traffic ceased in 1947 but the salt traffic continued until 1991. These two stations were close to each other and neither near Over, or Wharton for that matter. They were, however, much better placed for Winsford than the main-line station of that name.

Winsford and Over station is seen on the occasion of the RCTS railtour on 4 May 1952.

At Over and Wharton a LNW 1P 2-4-2T waits to depart.

Ingleton

The North Western Railway was incorporated on 26 July 1846 with powers to build a line from Skipton to Low Gill, on the L&C, plus a branch from Clapham to Lancaster. Once the LNWR was formed the NWR inevitably became known as the Little North Western. By 30 July 1849 the 30 miles from Skipton to Ingleton were completed and a service of four trains daily was put in place. However, once the NWR route to Lancaster was completed the passenger service on the 4-mile branch to Ingleton was withdrawn. The NWR no longer had the resources to complete the line to Low Gill and it was the L&C which completed the line, which opened in 1861. The L&C, now leased by the LNW, had its own station at Ingleton. It was the LNW's obstructive attitude towards the MR, now the lessees of the NWR, which caused the MR to plan its own route to Carlisle and Scotland. Once the Settle and Carlisle line was open the Ingleton line settled back to being a country backwater, instead of the through route to Scotland for which it had been intended.

In 1895 in the Down direction there were three services between Ingleton (MR) and Tebay and one to Low Gill only. All these Down trains had a long wait at Ingleton (LNW), presumably to carry out shunting. In the Up direction there were three services from Tebay to Ingleton (MR) and one to Ingleton (LNW). The LNW station at Ingleton closed in 1917. In 1953, the last year of passenger operation, there were six trains in each direction with an additional service on Saturdays. Passenger services ceased on 30 January 1954 and the line closed completely in 1965.

Right: Ingleton MR station.

Below: The viaduct connecting the two Ingleton stations. A train heads south.

Golborne

The Wigan Branch Railway opened its 6.9-mile line from Parkside, on the Liverpool & Manchester Railway, to Wigan Chapel Lane on 3 September 1832. Traffic from the outset was worked by the L&M. In 1843 the WBR merged with the Wigan & Preston Railway to form the North Union Railway. It was this company which opened Golborne station *c.* 1839. In 1895 it enjoyed a service of eleven trains a day on weekdays with just two on Sundays.

The service ran between Warrington and Wigan. In 1953 this had reduced to eight on weekdays, and on Sundays two Up trains and one Down. Renamed Golborne South in 1949, the station closed on 6 February 1961. In 2021 a sum of £15 million was agreed by the Mayor of Greater Manchester for the reopening of Golborne station. No date has been set.

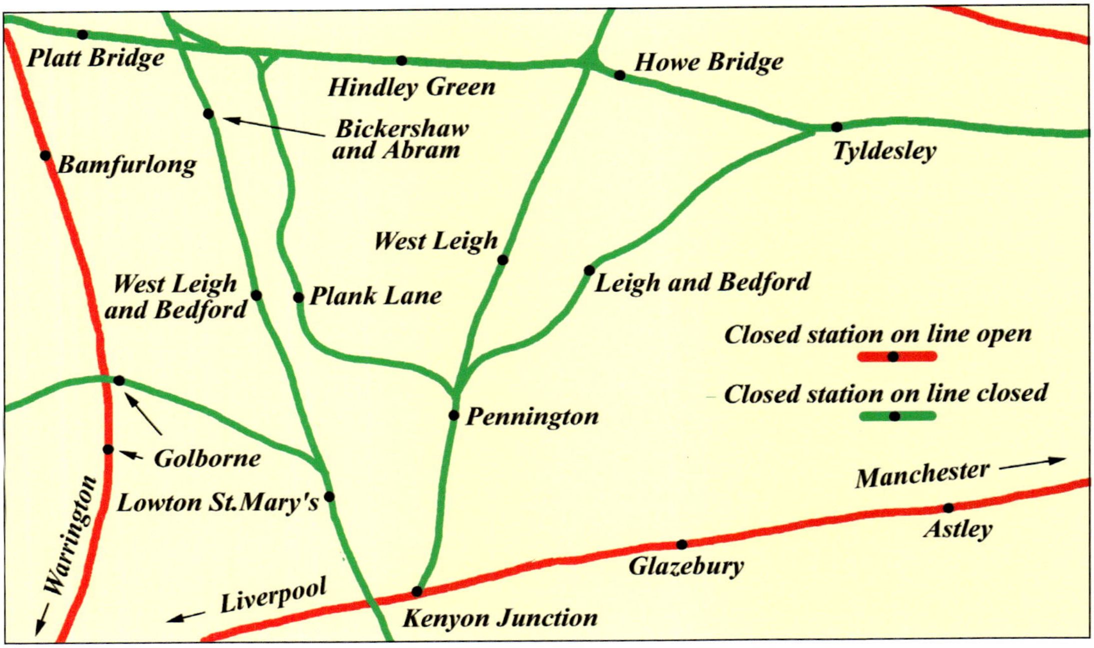

Railways in the area around Golborne and Leigh.

Golborne North station was on the GC Lowton St Mary's to St Helens branch. It opened for goods on 1 July 1895 but it was to be another five years before it saw passenger trains. This branch was intended to reach Southport, an ambition never realised. The station closed on 1 March 1952.

Golborne South station.

Leigh, West Leigh, West Leigh and Bedford, Atherleigh, Plank Lane, Pennington

This group of stations, irrespective of their names, all served the Leigh area. It seems implausible that an area once so well provided with stations should eventually find itself without a single station or even a railway line. The Bolton & Leigh Railway was the first public railway to open in Lancashire. The Act was passed on 31 March 1825 and goods traffic started on 1 August 1828. The main purpose of the railway was to connect Bolton with the Leeds and Liverpool, and Bridgwater Canals, which made an end-on connection at Leigh. On this 7¾-mile line there were rope-worked inclines at Daubhill and Chequerbent, where there were also stations, as well as at Atherton Bag Lane and West Leigh. The first locomotive was Stephenson's 0-4-0 *Lancashire Witch*. The line was extended to Kenyon Junction on the Liverpool & Manchester Railway in 1831. Passengers started being carried on the whole line from 11 June 1831. It was now possible for the first time to travel by train from Bolton to Liverpool and Manchester. Both the B&L and the Leigh & Kenyon Railway became part of the Grand Junction Railway in 1845. In 1935 a station was opened at Atherleigh. It closed with the rest of the line on 29 March 1954.

On 1 September 1864, a branch was opened from Tyldesley, on the Eccles–Wigan line, to Pennington. A station was opened at Bedford and Leigh, renamed Leigh and Bedford in 1876, and Leigh in 1914. Pennington to Bickershaw Junction was opened for goods on 9 March 1885, and a passenger service commenced with the opening of Plank Lane on 1 October 1903. Plank Lane station closed on 20 February 1915. Train services on the Eccles–Wigan line from Manchester to Bolton and to Wigan were withdrawn during the Second World War. The railways in the area around Leigh closed progressively to freight leaving just the lines serving Parsonage and Bickershaw collieries. Bickershaw Colliery, by this time combined with Parsonage, closed on 13 March 1992. Part of the Eccles–Wigan line and the branch from Tyldesley to Leigh is now a guided busway.

The Wigan Junction Railway was authorised by an Act of 16 July 1874. The line was constructed under the sponsorship of the MS&L and opened from Glazebrook to Strangeways Hall Colliery on 16 October 1879. Passenger services started after the line had been extended to a temporary terminus at Darlington Street Wigan in 1884. The original name for West Leigh and Bedford was Plank Lane for West Leigh. It was renamed in January 1894. It wasn't until 1906 that the WJR became vested in the GCR. In 1953 West Leigh and Bedford had a service of ten or eleven trains to and from Manchester Central or Irlam. There were five or six trains on Sundays. In 1962 there were just six trains a day on weekdays. All these trains ran to Manchester Central. Although there were Sunday trains on the line, none stopped at West Leigh and Bedford. A peculiarity from the 1953 timetable was retained, where the first train of the day from Wigan stopped at all the other stations on the line except West Leigh, except on Saturdays. Passenger services were withdrawn on 2 November 1964.

West Leigh station.

West Leigh and Bedford.

Work being undertaken in preparation for the opening of Atherleigh station.

Pennington station.

A National Coal Board Hunslet 'Austerity' 0-6-0ST hauls MGR wagons at Bickershaw Colliery in 1983. (Oxyman)

Mickle Trafford

The Birkenhead Joint line from Warrington to Chester was opened on 31 October 1850 but it wasn't until 1889 that a station was opened at Mickle Trafford. The CLC line for Northwich to Chester opened on 2 November 1874 and the station at Mickle Trafford opened the following year on 1 May. The CLC line ran close to the BJ line before veering away and passing over to reach Chester Northgate station. There was no connection between the two lines until the needs of the Second World War produced one in 1942. In 1895 the CLC station saw seven trains a day running Manchester to Chester, with two extended to Ellesmere on the Cambrian system. On the BJ line there were also seven trains in each direction to Chester, in this case from Manchester Exchange. The station also benefitted from the LNW service between Liverpool and Chester.

By the summer of 1947 each station was served by trains running between Chester and Manchester: just three or four at the BJ station, and five at the CLC station. These latter included trains to and from Llandudno. Neither station had a Sunday service. Both stations lost their passenger services on 2 April 1951. In 1969 a new connection was made to allow trains on the CLC line to run into Chester General.

Mickle Trafford CLC station.

Precursor 4-4-0 No. 25311 passes through Mickle Trafford Birkenhead Joint station in LMS days.

Cheadle

By an Act of 22 July 1861 the Stockport, Timperley & Altrincham Junction Railway was empowered to build a line from Portwood (Stockport) to Deansgate Junction. This was in fact an extension of the previously built Woodley–Portwood line, opened in 1863. The STAJ was opened on 1 December 1865, connection with the LNW at Broadheath was completed in 1866, and Cheadle station opened the same year. The station was served by trains running between Liverpool and Stockport on the CLC lines. This continued into BR days and in 1953 just four trains a day called on weekdays. Passenger services were withdrawn on 30 November 1964. Of the CLC line that once reached form Godley to Glazebrook, only the section from Cheadle Junction to Skelton Junction remains open, and currently sees freight traffic. Cheadle station buildings now house a restaurant.

The line from Edgeley Junction (Stockport) to Northenden Junction, where it joined the STAJ, was opened by the LNW on 1 August 1866. It used the CLC line to reach Broadheath and thence to Warrington. Cheadle station closed on 1 January 1917 but the line remained open for freight traffic. On 15 May 1989, it started to see passenger trains again when the Manchester to Chester trains were diverted via Stockport and Skelton Junction. This was caused by the conversion of the Manchester–Altrincham railway to a tram line. Discussions are currently ongoing with a view to opening a new Cheadle station on the LNW line.

A Sentinel steam railcar at Cheadle North.

Northenden Junction was the meeting point of the CLC and LNW lines just to the west of the two Cheadle stations. On 11 May 1953, a Class 142 'Pacer' passes with a Chester service while in the background Class 37 No. 37358 waits with a Tunstead–Northwich limestone train. On the right is the Greater Manchester waste depot.

The East Midlands

Tuxford

The LD&ECR station at Tuxford opened with the line on 15 December 1896. The railway became part of the GCR in 1907 and in 1923 the station had 'Central' added to its name. To the east of the station there was a west to north chord joining the LD&ECR line to the Great Northern main line. Within the triangle made by the chord was Tuxford locoshed. To the north of the running lines was Tuxford Locomotive Works, more familiarly known as 'The Plant'. In 1955 the station, by now Tuxford West, saw just three trains daily in each direction. It closed on 19 September that year, along with the other stations on the line. The locoshed closed in 1959. The line was retained for goods traffic, particularly for coal, as there were a number of collieries along its route. Trains

In the early 1950s Robinson C13 4-4-2T pauses at Tuxford Central with a service for Chesterfield.

ran to High Marnham Power Station, situated near Fledborough station. The power station ceased operations in 2003. However, the 10-mile-long line has been retained and is now the Tuxford Rail Innovation and Development Centre, run by Network Rail.

Where the LD&ECR line passed over the GNR main line the two companies built an exchange station with platforms at both levels. Originally Tuxford Exchange, the name was later changed to Dukeries Junction. By 1947 there was just a single daily service stopping at the low-level platforms and the station closed altogether in 1950. The GNR also had a Tuxford station, situated to the north of Dukeries Junction. It opened in 1852, was renamed Tuxford North in 1923, and closed in 1955.

Above: Tuxford Dukeries station.

Right: Tuxford North.

Above: At Tuxford locoshed a line-up of Robinson 2-8-0s together with a J11 0-6-0.

Left: On 25 March 1993, to the east of the former Tuxford Central station Class 58 No. 58037 heads west with an empty MGR from High Marnham Power Station.

Haxey

After two previous attempts, the line between Gainsborough and Black Carr Junction, on the GNR main line, was finally authorised in 1864 and opened on 15 July 1867. Haxey was one of the original stations. It was later renamed Haxey and Epworth. The GER had for a long time wished to penetrate GNR territory, not least to participate in the lucrative Yorkshire coal traffic. Finally, in 1879 an agreement was reached to operate a joint line from Huntingdon to Doncaster, via March and Lincoln. The new joint line incorporated sections of railway that had already been built, including the line through Haxey. A new section of line between Spalding and Pyewipe Junction was needed to complete the line and when this opened in August 1882 the GN/GE joint committee took over the management of the line. In 1955 Haxey had just four trains calling on weekdays and two on Sundays. It lost these services in 1959.

The Goole & Marshland Light Railway was authorised in 1898 to build a line from Marshland Junction, on the Doncaster–Goole line, to Adlingfleet via Easton, with branches to Swinefleet and Luddington. In the following year the Isle of Axholme Light Railway was authorised to build a line from Haxey, where it would connect with the GN&GEJt, to Reedness Junction, where it would connect with the G&MLR. Construction was under way when both lines were purchased jointly by the NER and the L&Y, becoming the Isle of Axholme Joint Railway. It opened on 2 January 1905.

Haxey Junction station was alongside Haxey and Epworth station. Further north was Haxey Town, originally named Haxey Central. The latter title was probably justified as the IALR station was considerably nearer to the village than the GCR. Trains were worked by the L&Y, mainly with Barton Wright 0-6-2Ts and Aspinall 0-6-0s and 2-4-2Ts. One of the 0-6-2Ts is seen at Haxey Junction. These engines, introduced in 1880, were the first of that wheel arrangement in Britain. The LMS introduced a Sentinel rail car in 1926 and a larger one in 1930, which saw out the remaining years of the passenger service, which ended in 1933. Freight continued until 1965.

Haxey and Epworth station.

Haxey Town station.

L&Y Barton Wright 0-6-2T No. 229 stands with its train at Haxey Junction.

The Welsh Borders

Brymbo and Plas Power

The first railway to reach Brymbo was the branch to Minera from Wheatsheaf Junction on the North Wales Mineral Railway Ruabon–Saltney line. This opened in July 1847. It included two inclined planes and when the Wrexham & Minera Railway opened in 1862 the Wheatsheaf line saw little further use. The W&M ran from Croes Newydd to Brymbo where it met the earlier line. In 1872 this line was extended 2¾ miles to meet the LNW at an end-on junction between Llanfynydd and Coed Talon. Passenger trains between Wrexham and Brymbo began in 1882 and between Mold and Brymbo in 1898. The attractions for the railways at Brymbo were the steelworks and collieries, and passenger traffic was of little importance. In 1902 there were just four trains a day between Wrexham and Coed Poeth (next station after Brymbo), with extra trains on Thursdays and Saturdays. A similar service operated between Mold and Brymbo. The service between Wrexham and Berwig (opened 1905) lasted only until 1 January 1931 while the Mold–Brymbo trains ended in 1950, by which time the service was down to two trains daily. Freight traffic continued until 1982.

In 1882 the Wrexham, Mold & Connah's Quay Railway was authorised to build a line to Brymbo from Gweersyllt, on the Wrexham–Connah's Quay line. The line was opened throughout by 1887 and passenger services started on 1 August 1889, with stations at Moss and Pentre, Plas Power and Brymbo. A number of halts opened later. The WM&CQ became part of the GCR in 1901. In 1895 the passenger service from Wrexham Central consisted of just three trains daily but on Saturdays there were no less than twelve trains. This service ceased altogether on 1 March 1917 while freight continued until complete closure of the line in 1970.

The GWR station at Plas Power opened in February 1883 and was served by the Wrexham–Coed Poeth service. The station closed in 1931. The WM&CQR station was adjacent to that of the GWR and opened on 1 August 1889. The passenger service lasted only until 1917.

The GWR station at Brymbo, long out of use when photographed in 1962. (Ben Brooksbank)

The WM&CQ 0-6-2ST No. 15 depicted here at Brymbo was built by Beyer-Peacock in 1882 and survived in service until 1922.

A view looking south from Brymbo Middle signal box. The lines to the left lead to the GWR station, while those straight on lead into the steelworks. (Ben Brooksbank)

Brymbo West Crossing Halt was among a number of new halts opened by the GWR in 1905. It turned out to be better located for many people than Brymbo itself but closed in 1931 when the passenger service was withdrawn. In 1962 at the former location of the halt a Class 56XX approaches with a Down freight. (Ben Brooksbank)

The crew of a GWR 0-6-0ST and station staff pose for a photograph at Plas Power GWR station.

Oswestry, Whittington and LLanymynech

The Shrewsbury & Chester Railway opened throughout on 12 October 1848. Whittington station opened on the same date. In 1902 it had a somewhat unbalanced service pattern. In the Down direction there were seven trains on weekdays including two to Liverpool, one to Wrexham, and one to Chester. In the Up direction there were four trains to Shrewsbury and one to Birmingham. The first train in the Up direction did not leave until 09.40 and the second not until 12.33. Not much had changed by 1958; there were now five services daily with trains running mostly between Shrewsbury and Chester. There were three trains on Sundays. 'Low Level' was added in 1924.

Oswestry joined the railway network when the 2¼-mile branch from Gobowen was opened by the S&C on 23 December 1848. The line between Oswestry and Newtown opened throughout on 10 June 1861. The Oswestry, Ellesmere & Whitchurch Railway opened to passengers on 27 July 1864. The line from Whitchurch to Newtown was to become the main line of the future Cambrian Railways.

The Shropshire & Montgomeryshire started life as the Shrewsbury & North Wales Railway. The line from Abbey station, Shrewsbury, to Llanymynech, together with two branches, opened on 13 August 1866. Before that time the S&NW had amalgamated with the Shrewsbury & Potteries Railway and was now known as the Potteries, Shrewsbury & North Wales Railway. In only one of the succeeding years was a profit made. In 1880 the railway was in such a poor state that it was forced to close by the Board of Trade. In 1890 an attempt to reopen the line was made by a new company but failed due to lack of finance. In 1911 Colonel Stephens arrived on the scene and reopened the line as a light railway. The Potts, as it was familiarly known, became the Shropshire & Montgomeryshire Light Railway. The passenger service ceased in 1933 and in 1939 the line came under military control, a number of the stations being used as ammunition dumps. Final closure came on 29 February 1960.

Whittington Low Level.

Above: The station closed on 12 September 1960 but the station building still stands, as seen in this May 1991 view. It is now a private dwelling.

Right: The lever frame of Whittington signal box. The box dates from 1881 and the lever frame from 1912. The box was abolished in 1992 when automatic barriers were installed.

Left: Whittington High Level is seen being inspected by staff following a fire in 1958. The station closed on 4 January 1960.

Below: Oswestry seen from the north in 1962. The building on the right-hand side is the former GW station. Following the amalgamation of the Cambrian and GWR, from 1924 the latter diverted all its services to the Cambrian station, the old station being used as a goods depot.

On 12 April 1960 ex-GW 'Manor' 4-6-0 No. 7815 *Fritwell Manor* stands at the Up platform at Oswestry. No. 7815 was a Machynlleth locomotive at this time. (Ben Brooksbank)

On 21 August 1957 ex-GW 0-6-0PT No. 5405 departs from Oswestry with a service for Gobowen. No. 5405 was built in 1931 and spent its whole working life stationed at Oswestry. It was withdrawn just two months after this photograph was taken. After the Cambrian main line was closed in 1965 a DMU shuttle continued between Gobowen and Oswestry until 7 November 1966. (Peter Groom)

Freight services at Oswestry were withdrawn in 1971 but trains continued to pass en route to Llanyblodwell Quarry until 1988. In 1991 the track remains in good condition, although trains have ceased. The large building on the right is the former headquarters of the Cambrian Railways. It also served as the station building. Cambrian Heritage Railways is now based at Oswestry with the eventual aim of restoring a tourist railway service between Gobowen and Blodwel.

A Cambrian Railways train is seen at LLanymynech. The locomotive is Sharp Stewart 61 Class 4-4-0 No. 65 built in 1893. At the amalgamation of the CR and GWR this engine was allocated the number 1096 but never actually carried it, being withdrawn in 1922.

In 1962 an Ivatt 2MT 2-6-0 arrives at Llanymynech with a southbound service. The Shropshire & Montgomeryshire Railway station was to the right. (Lamberhurst)

Llanymynech S&M station in 1903. The name of the station can just be made out on the front of the canopy.

Among the Colonel's eclectic mix of motive power and rolling stock was this Ford lorry and trailer.

Scotland

Kilbirnie, Beith, Lugton, Uplawmoor

An Act of 28 July 1884 authorised the 4-mile, fifteen-chain branch to Kilbirnie from the Barrmill & Kilwinning Railway, which had been authorised by an earlier Act. The main purpose of the branch was to serve the Glengarnock Steel Works, where there was also a station. The branch opened for goods on 1 November 1889 and to passengers a month later. Following the Grouping the station became Kilbirnie South. The station had a service of eight or nine trains daily with additional services on Saturdays but none on Sundays. The branch closed to passengers and freight on 1 December 1930.

In response to congestion on its main route to Ayr and Kilmarnock, in 1905 the G&SW constructed a new line from Johnstone North to Dalry. Through Dalry there was quadruple track until the routes diverged to Ayr and Kilmarnock. Kilbirnie was one of four new stations on the line. There was already a Kilbirnie station on the original main line. This had become Glengarnock and Kilbirnie in 1888 and with the opening of the new line it became simply Glengarnock. In 1961 the line still had a reasonable service with some fourteen trains on weekdays, partly provided by DMUs. Clearly, closing this parallel route would make considerable savings and so it was that in 1966 passenger services were withdrawn. Freight traffic continued until 1972, and until 1977 the line from Brownhill Junction (north of Dalry) remained open, primarily to serve Glengarnock Steel Works.

The Glasgow, Paisley, Kilmarnock & Ayr Railway (later G&SW) opened from Kilwinning to Beith on 21 July 1840 and from Beith to Paisley in stages up to 12 August 1840. The line from Paisley to Glasgow, joint with the Glasgow, Paisley & Greenock (later Caledonian), was opened on 13 July the same year. Although other stations have remained open, Beith North, as it became after the Grouping, closed on 4 June 1951. Its demise was probably partly due to its distance from the town, Beith Town being much better placed. The Town station was the terminus of a branch which left the main line of the Glasgow, Barrhead & Kilmarnock Joint Railway at Lugton. Both the main line and the branch opened on 26 June 1873. The railway was joint between the GSWR and the Caledonian but after January 1917 the Caledonian ceased running passenger trains and these were left in the hands of the GSWR. In 1961 Beith Town enjoyed a service of fifteen trains daily in each direction, one of which in both the morning and evening peak was direct to/from St Enoch. The others ran to and from Lugton. Passenger services were withdrawn on 5 November 1962. Freight continued for a further two years.

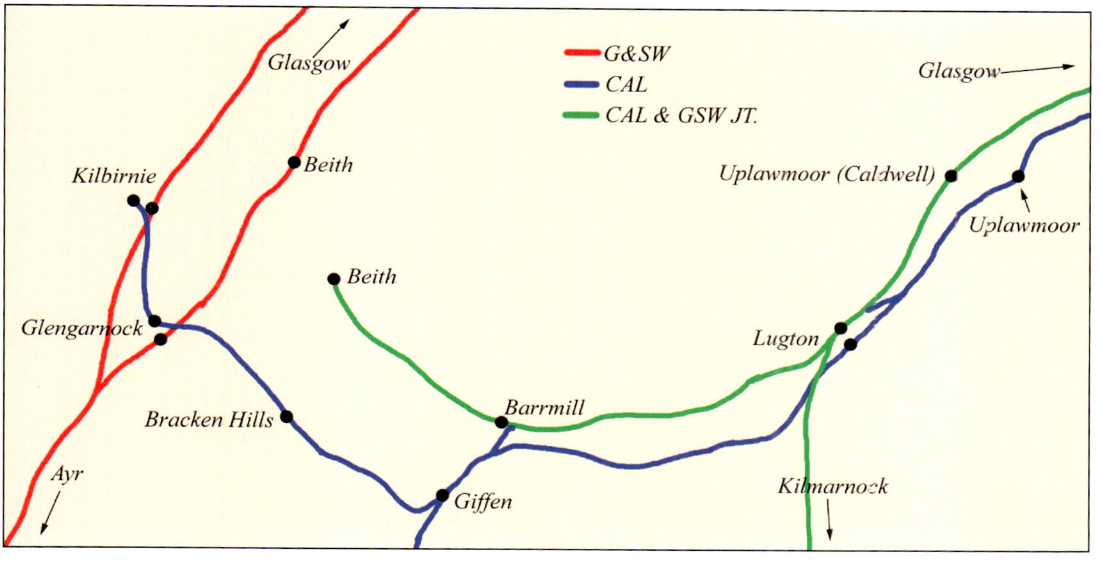

Kilbirnie, Beith area.

Kilbirnie Caledonian station after closure.

Kilbirnie GSW, looking towards Glasgow.

Bristol railcar No. Sc79558 is seen at Beith with the Lugton shuttle in 1959. This unit was shortly replaced by an AC railcar.

At Lugton Junction in 1966 photographers are waiting for something special. The station lost its passenger service on 7 November that year.

Uplawmoor, the next station towards Glasgow, lost its passenger trains on the same date. From its opening in 1871 the station was known as Caldwell. It only received the name Uplawmoor with the closure of the Lanarkshire & Ayrshire Railway station of that name in 1962. It only had four years to enjoy its new name, closing on the same date as Lugton Junction. This is the L&A station, looking towards Glasgow.

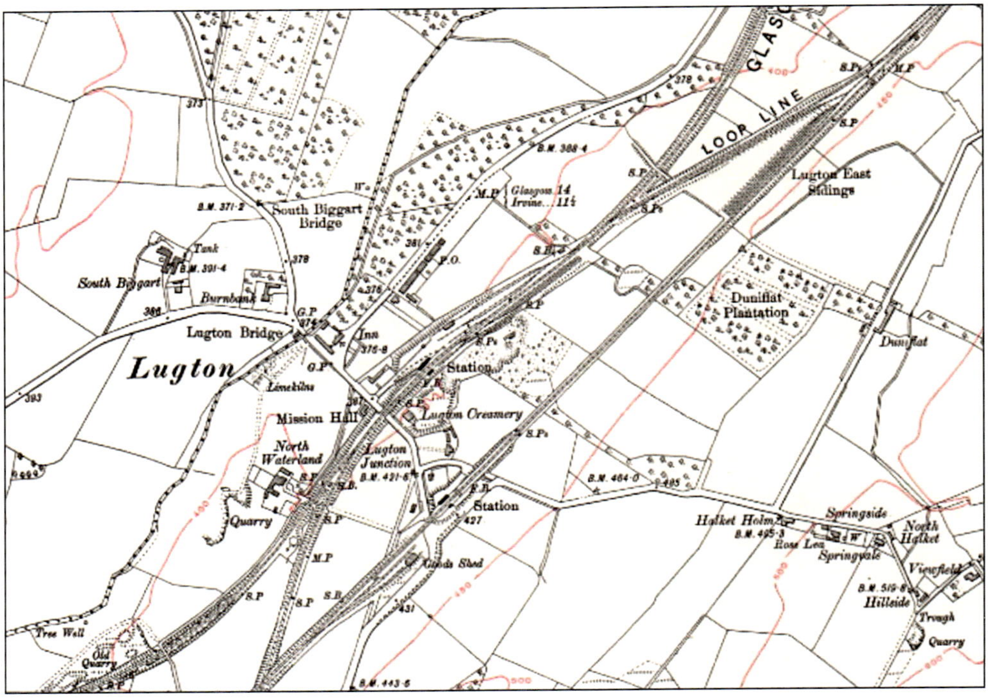

Lugton High station was also on the Lanarkshire & Ayrshire Railway. An Act of 1897 authorised a line from Newton to Lugton and on to Giffen Junction on the original L&A line. This line had opened between Barrmill Junction, on the GB&K Beith branch, to Ardrossan in 1888. The main purpose of the line was the transport of coal from the Lanarkshire coalfields. The line opened for goods traffic on 6 January 1904. The passenger service beyond Uplawmoor was withdrawn in 1932. Today the L&A remains open only as far as Neilston. The closeness of the two Lugton stations is clear from the map.

Buckie, Portessie

On 12 July 1882 an Act was passed enabling the Highland Railway to build a line from Keith to Portessie. On the same date the Great North of Scotland Railway was authorised to extend their line from Grange to Portsoy along the coast to Elgin. The 13½-mile HR branch was completed first, opening on 1 August 1884. There were just four services each way on weekdays only. Some of these were mixed trains and journey time for the length of the line could be as much as fifty minutes. Services were suspended in 1915 and the track lifted for use elsewhere. It was relaid after the Grouping but passenger services were never reinstated. The line between Keith and Aultmore remained open until 1966 for trains to the distillery at the latter location, and also between Portessie and Buckie station where fish was loaded for LNER services up until 1944. The GNoS line was opened in stages between 1 April 1884 and 1 August 1886. The line was single track except for the section between Buckie and Portressie. In 1895 there were five trains daily in each direction on weekdays only. There

An early photograph of fish sellers at Buckie HR station.

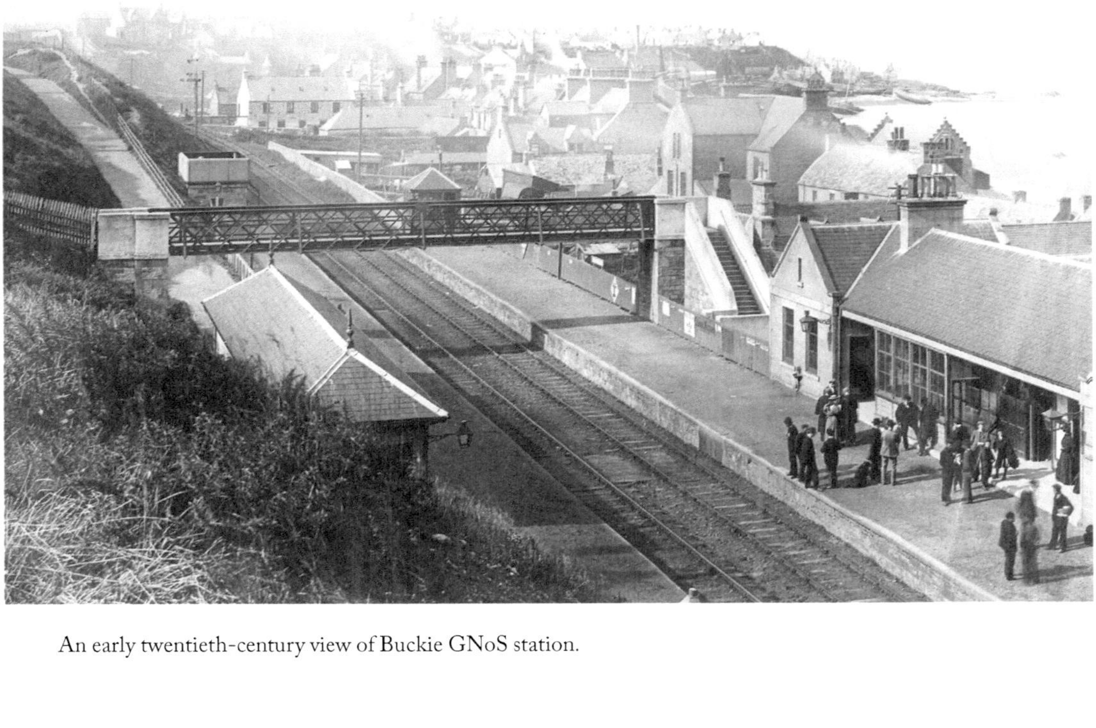

An early twentieth-century view of Buckie GNoS station.

This is Portessie in 1954. The HR trains used the outer face of the platform on the right. Some of the HR facilities can just be glimpsed on the right of the photograph. In 2021 the station building and platform still stood, together with the turntable pit and water tower. However, they were about to disappear under a housing development.

were in addition three semi-fasts which stopped at Buckie but not at Portessie. In 1961 DMUs were introduced to the line with six trains daily. Most ran to/from Aberdeen. Closure came to the whole line on 6 May 1968.

Newport-on-Tay

The two Newport-on-Tay stations opened on 13 May 1879, following the opening of the Tay Bridge. The Tay Bridge Disaster of December the same year curtailed services for the next eight years, until the new Tay Bridge was opened. In 1895 there were thirteen services daily from Dundee to Tayport, four to St Andrews, three to Leuchars, and two to Newport East. On Sundays threre were six services to Leuchars or Tayport. In 1956 there were still twenty-two services daily. In 1966 services were cut back to Newport East due to the new Tay Road Bridge being

On 16 March 1967 B1 4-6-0 No. 61278 approaches Newport West with a single coal wagon.

Newport East had a passing loop until the railway was severed to the east of the station in 1966, after which it was lifted. Passenger services were provided by these Class 101 units. Notice the two coal merchants in the background and the venerable old carriage used as an office.

built across the railway line east of Newport East. The bridge opened in 1968 and sounded the death knell of the railway as it was easier for people to drive or take a bus across the bridge to Dundee. Both stations closed on 3 May 1969.

Grantown-on-Spey

The Inverness & Aberdeen Junction Railway was less than happy that its outlet to the south was via Aberdeen and the GNoS. A new main line was authorised on 22 July 1861 to run from Forres to Dunkeld. This was the Inverness & Perth Junction Railway. Perth to Dunkeld had been completed in 1856. This 119-mile line was completed in just under two years. The initial service was of just one train a day taking 6 hours between Perth and Inverness. In 1865 the I&P and the I&AJ merged to form the Highland Railway. By 1885 the fastest train took 4 hours. From 1898 trains between Perth and Inverness used the new line via Carr Bridge and the Speyside line was reduced to a local service. In 1953 there were four or five trains daily calling at Grantown, one of these on Saturdays was direct to/from Glasgow. The line from Aviemore to Forres closed on 18 October 1965.

The Strathspey Railway was authorised by an Act of 17 May 1861 to build a line from Dufftown via Craigallachie to Boat of Garten, where it would meet the Highland Railway. This 33¼-mile single-track line was opened as far as Abernethy (later Nethy Bridge) on 1 July 1863. Thus the lucky inhabitants of Grantown now had two stations opened in the same year. Trains did not run to Boat of Garten until 1866. In that same year the railway became part of the GNoS. For most of its life the line saw just three return trains daily. Wickham railbus No. 79966 was sent to work the line in 1959. The service provided by these railbuses at this time consisted of three trains in between Aviemore and either Keith, Elgin, or Craigellachie. There was an additional service on Saturdays and none on Sundays. Passenger services were withdrawn on the same day as at Grantown on Spey West. Freight continued for a few more years.

Above: Grantown-on-Spey West seen some time after closure.

Left: The railbus is seen at Grantown East with a southbound service.

Renfrew

The Paisley & Renfrew Railway opened between Hamilton Street, Paisley, and Renfrew Wharf on 3 April 1837. The line was built to a gauge of 4 feet 6 inches and the three locomotives provided initially were withdrawn in favour of horse haulage from March 1842. The locomotives were later sold at public auction. In 1847 the P&R was taken over by the Glasgow, Paisley, Kilmarnock & Ayr Railway (later GSWR). In 1866 the line was doubled and converted to standard gauge and a steam service introduced between Glasgow and Renfrew Wharf. Fulbar Street was opened on 1 May 1866 and Renfrew South on 1 May 1897. The Renfrew & District Railway, part of the Glasgow and Paisley Joint Line, was opened to passengers on 1 June 1903. There were stations at Deanside, Kings Inch (originally Renfrew Central), and Porterfield, the latter being alongside South Renfrew station. The Caledonian and GSWR provided passenger services in alternating six-month periods. In 1908 there were seven or eight trains daily to King's Inch and Porterfield. Deanside had already closed. In fact it was difficult to see why it had ever opened, located in the middle of nowhere. There was a similar service to Renfrew Wharf. Passenger services on the RDR

At Renfrew Wharf McIntosh 812 Class 0-6-0 No. 57566 waits to depart. The lines on the left lead into the yard of Babcock and Wilcox. The chimneys in the background belong to Yoker Power Station. (Neville Stead)

Passing South Renfrew No. 2 signal box on 15 August 1958 is a Standard Class 3MT 2-6-0 with the 17.15 service from Renfrew Wharf to Glasgow St Enoch. (BLS)

ceased altogether in 1926, but the line remained open for freight traffic. Part of the line as far as Deanside is still open today to serve the freight depot of J. G. Russell. The line to Renfrew Wharf closed to passengers on 5 June 1967.

Granton, Leith

The first railway in Edinburgh and also to Leith was the Edinburgh & Dalkeith Railway. Various Acts up to 1834 authorised the construction of a line from St Leonards to the River South Esk near Dalhousie, and branches from Niddrie to Leith and Fisherrow. The 3½-mile Leith Branch opened in 1838. These lines were built to a gauge of 4 feet 6 inches and traction was by horse. The primary purpose of the railway was the transportation of coal to Edinburgh but a passenger service proved surprisingly popular. There were no stations and the carriages would stop wherever there was a passenger. In 1845 the railway was acquired by the NBR for £113,000. The line was rebuilt to standard gauge and steam traction employed from 1858. South Leith closed to passengers on 2 January 1905. A line to the docks at Leith still exists but it is little used and its future is uncertain.

The Edinburgh, Leith & Newhaven Railway was authorised in 1836. An Act of 1839 superseded the original and the aims were changed to building a railway from Scotland Street to Trinity. A further Act of 1844 changed the name of the company to the Edinburgh, Leith & Granton Railway and authorised branches to the two last named locations. The branch to Granton opened on 19 February 1846, with the branch to North Leith following on 10 May the same year. Before the construction of the Forth Bridge, passengers would take the ferry from Granton to Burntisland, where they could continue their journey by rail. The line between Scotland Street and the terminus at Canal Street was completed in 1847. It included a tunnel 1,052 yards long with a gradient of 1 in 27. It was worked by cable. At this time there were twenty-five services daily between Canal Street and Granton. In 1848 the EL&G was absorbed by the Edinburgh & Northern Railway, which was renamed the Edinburgh, Perth & Dundee Railway from April 1849. It became part of the NBR on 29 July 1862. To avoid the severe gradient in the Scotland Street Tunnel the NBR built a new line which ran via Abbeyhill to connect with the Granton and Leith branches. This was opened in 1868. The station of Leith Walk was on this line. Granton lost its passenger service on 2 November 1925. The final timetable provided a service of eight or nine trains daily. By 1939 there were eleven trains daily between North Leith and Edinburgh Waverley, several fewer on Saturdays. This service was withdrawn in 1947. Freight continued until 1986.

In 1903 the NBR constructed a line from a junction near Abbeyhill station to a new station, Leith Central. This was a building on a grand scale with four platforms, seldom all used. In the first year of Nationalisation the trains emanating from Central were mainly Circle Line services, with a few running to Waverley only. This came to an end in 1952, after which until 1972 the station was used as a diesel maintenance depot and driving school. The building still stands.

The Caledonian opened a freight line to Granton Harbour in 1861. From this line a branch to Leith Western Docks was opened in September 1864. From Newhaven a branch to Leith (North) was opened on 1 August 1879 and a service of twelve trains daily to Princes Street started. By the 1930s this had grown to forty trains a day, hauled mainly by 0-4-4 tanks. The service was withdrawn on 20 August 1962, thus being the last Leith station to close to passengers.

The Leith New Lines was a project of the Caledonian to build a circular route on the north side of Edinburgh. The only section authorised was the line from Newhaven to Leith Eastern Docks, which opened in 1903. Stations were built at Ferry Road and Leith Walk but were never opened and the line was only ever used for freight.

At Granton NBR station Class R 4-4-0T (LNER Class D51) No. 78 stands ready to leave with a train for Edinburgh.

There was also a station at Granton Gasworks. Intended for employees at the works, it was also used by ordinary members of the public. This opened on 27 November 1903 with trains being provided by the CR. It closed in 1942. The station is seen on the opening day with McIntosh 0-6-0 No. 821 in attendance.

Leith North, the CR station, seen in LMS days.

At North Leith is D51 No. 10467 seen in LNER days.

The opening day at the new NBR station of Leith Central. On 1 July 1903, NBR 4-4-0T No. 33 stands ready to leave for Edinburgh with the inaugural train.

Leith Central today.

Bothwell

The NBR line to Hamilton branched off the Glasgow–Coatbridge line at Shettleston Junction. It was built by the Glasgow, Bothwell, Hamilton & Coatbridge Railway, incorporated by an Act of 16 July 1874. In fact two lines were built, the second leaving the Coatbridge line at Sunnyside Junction. The two lines met at Bothwell Junction. The station at Bothwell opened in 1878, the same year the GBH&C became part of the NBR. The GBH&C owned four 0-6-0Ts built by Dubs and Company in 1877, and later rebuilt by the NBR. Between 1917 and 1919 the line lost its passenger service as a wartime economy measure. In 1952 Bothwell became a terminus when the line beyond to Hamilton closed. Passenger services were withdrawn in 1955 while freight continued until 1961.

The Caledonian station was the terminus of a branch from the Glasgow–Carstairs main line. It opened in 1877. In 1895 it had a service of nine or eleven trains daily, with additional services on Saturdays. By 1948 this had declined to three services in the Down direction and four in the Up. There was no Up train after 1.12 p.m. The line closed completely two years later.

At Bothwell NBR in the early 1950s N2 0-6-2T waits to depart.

Peebles, Dolphinton

Several attempts were made to bring the railway to Peebles in the mid-nineteenth century. None succeeded until the Peebles Railway obtained an Act of 8 July 1853 authorising the construction of a branch of 18¾ miles from Hardengreen on the Hawick Branch of the NBR. This single-track line was opened on 4 July 1855 and worked by the NBR from the outset. The PR was absorbed by the NBR in 1876. In 1864 the railway was extended a further 18 miles to Galashiels. In that same year a line from Symington, on the Caledonian main line, reached Peebles. This station was on the south side of the Tweed, the PR station being on the northern side. A link between the two railways was completed the same year. The line was worked by the Caledonian, which absorbed the local company – the Symington, Biggar & Broughton Railway – even before the railway was completed. This line lost its passenger service on 5 June 1950, although freight continued for another four years. On the NBR line in 1895 there were five services on weekdays in both directions. By 1955 this had declined to three Up and four Down trains. However, in 1958 diesel units were introduced to the line with a much improved service of eight Up trains and six Down. All these services were between Edinburgh Waverley and Galashiels. However, this was not enough to save the line, which closed to all traffic on 5 February 1962.

The Leadburn, Linton & Dolphinton Railway was constructed on the Cheap Railway Principle, the chief advocate of which was Thomas Bouch, the engineer of the railway. Work started on the 10-mile line in March 1863 and the line opened on 4 July 1804. There were stations at Lamancha, Macbie Hill, Broomlee, and Dolphinton. Broomlee was the station for West Linton, the principal settlement on the line. There was an initial service of four trains daily, worked by the NBR, running between Leadburn and Dolphinton. By 1865 the line was in trouble with debts of more than £28,000 and absorption by the NBR was inevitable. This was authorised by an Act of 16 July 1866. By 1887 there were five trains daily in each direction, one of which was mixed. By the 1930s the line was not doing well, local bus services giving direct access to Glasgow having taken away much of the passenger traffic. It was initially thought that the line might stay open for freight but in the end it was closed to all traffic on 1 April 1933. The track was lifted between Dolphinton and Linton in 1936. In 1940 the remaining part of the line was used to store ammunition in more than 100 huts strung along the line. This continued until December 1946. Some freight continued until 1960 and in 1961 the remaining track was lifted.

Not to be outdone the Caledonian proposed a line from Carstairs to Dolphinton, with stations at Coulter, Biggar, Broughton, Stobo, Lyne, and Dolphinton. This line of 10 miles, fifty-nine chains was completed by February 1867. At Dolphinton the Caledonian had its own separate station. There was a connecting line between the two stations but it was only ever used for the passage of freight wagons. Faring little better than the NBR, line closure came on 12 September 1932. Protest caused the line to be reopened, until final closure in June 1945. Freight continued until 1950. Surprising perhaps that two lines were built to a tiny settlement in the middle of nowhere but perhaps even more surprising that a third line was once envisaged for the tiny hamlet. This would have been a joint GSWR/NBR line from Douglas. Fortunately wiser heads prevailed and the proposal was dropped.

Above: In the early years of the twentieth century Class M 4-4-0 No. 487 (LNER Class D31) waits to leave Peebles NBR with a service for Galashiels.

Right: At the Caledonian Peebles West station a McIntosh 0-6-0 waits to depart.

Dolphinton NBR.

The Caledonian station at Dolphinton.

Gretna

There were three stations named Gretna, two of them actually in England. The Gretna station in Scotland was on the Glasgow & South Western Railway line from Dumfries to Gretna Junction. The line was actually built by the Glasgow, Dumfries & Carlisle Railway, later part of the GSWR, and opened on 23 August 1848. It was renamed Gretna Green in April 1852 and closed on 6 December 1965. It was reopened on a new site on 20 September 1993 with just a single platform. A second was added later when the line from Annan was doubled.

The station on the Caledonian main line from Beattock opened with the railway on 9 September 1847. It lost its passenger service in 1951. Alongside the Caledonian station was Gretna NBR. This station was at the end of a branch, which left the Waverley line at Longtown and opened on 1 November 1861. In 1895 there was a service of just two trains daily, both in the morning, which ran from the GSW station via Gretna NBR to Longtown. This service was lost in 1915 when the line was requisitioned by the Ministry of War. A vast armaments depot was set up straddling the England/Scotland border stretching from Longtown to Eastriggs, serviced by an 80-mile-long railway network. This network had sixteen stations, including Gretna Township. There was also a 2-foot-gauge internal network some 125 miles long worked with battery locomotives.

The Caledonian station of Gretna Green.

The NBR Gretna Green station.

Above: LMS 4-4-0 Compound No. 1180 heads through Gretna Green GSWR with a westbound freight.

Left: Gretna Green Township station.

Strathaven

The Hamilton & Strathaven Railway was authorised by an Act of 10 August 1857. Its chief promoters were the ironmasters and colliery owners in the district around Quarter. It opened to that location on 9 August 1860 and to Strathaven on 2 February 1863. The station at Strathaven was at Fremington to the north-east of the town. It was worked by the Caledonian and absorbed by that company in 1864. For the next forty years this was the only railway at Strathaven. The 1895 timetable shows a service of seven trains daily in each direction with additional services on Saturdays. When the Stonehouse to Strathaven line opened in 1904 the H&S line was extended to meet it. A new station, Strathaven North, was opened on the extension, and Flemington became a goods station.

The new Caledonian lines stretched from Lesmehagon to Stonehouse, and from there on to Strathaven and Darvel. Passenger services began on 1 July 1905. These trains ran from Hamilton on the Quarter line, stopped at both Strathaven North and the new Strathaven Central station before continuing to Darvel. The services were provided on alternate six-month periods by the GSW and Caledonian. There was also a shuttle service between Stonehouse and Strathaven Central. The line was closed during the First World War and part lifted for use elsewhere. By 1922 only the section as far as Ryeland had been relaid and services reopened as far as here in October of that year. The line between Strathaven and Darvel closed in 1939. The line north to Hamilton closed to passengers in 1945 and completely in 1953. Services continued via Stonehouse until 4 October 1965.

Strathaven Central. Notice the driver of the McIntosh 0-4-4T standing on the buffer beam with his oilcan.

Strathaven North in LMS days.

Bonnybridge, Greenhill

Scotland's first railway between its two major cities opened between Glasgow Queen Street and Edinburgh Haymarket on 19 February 1842. The Edinburgh & Glasgow Railway had been authorised to build its 46-mile line in 1838. Bonnybridge station did not open until 1 May 1870. It was renamed Bonnybridge High in 1953, reverted to Bonnybridge in 1965 and closed on 6 March 1967. The Scottish Central Railway was formed to provide a route between Stirling and Perth, thereby connecting the Central Belt with northern Scotland. An Act was passed on 1 July 1845 for a 47-mile line from Greenhill, where the SCR would eventually make connections with both the E&G and the Caledonian, and Perth. The line was opened to the public on 22 May 1848. The SCR was amalgamated with the Caledonian in 1865. Last on the scene was the Kilsyth & Bonnybridge Railway, a joint venture of the NBR and Caledonian. It ran from Kilsyth Junction on the Kelvin Valley Branch of the NBR to Bonnywater on the Caledonian Denny Branch. In its 8½ miles there were stations at Kilsyth (New), Colzium, Banknock, Dennyloanhead, and Bonnybridge Central. The railway opened on 2 July 1888. There was a service of five trains daily running the length of the line, plus one daily to and from Queen Street. The railway remained nominally independent until the grouping, after which it became the joint responsibility of the LMS and LNER. The Central station was the best located of the three Bonnybridge stations but the passenger service was still withdrawn as early as March 1935. Freight continued on part of the line until 1964. The E&G and SCR lines remain open but there is no longer a station at Bonnybridge or Greenhill.

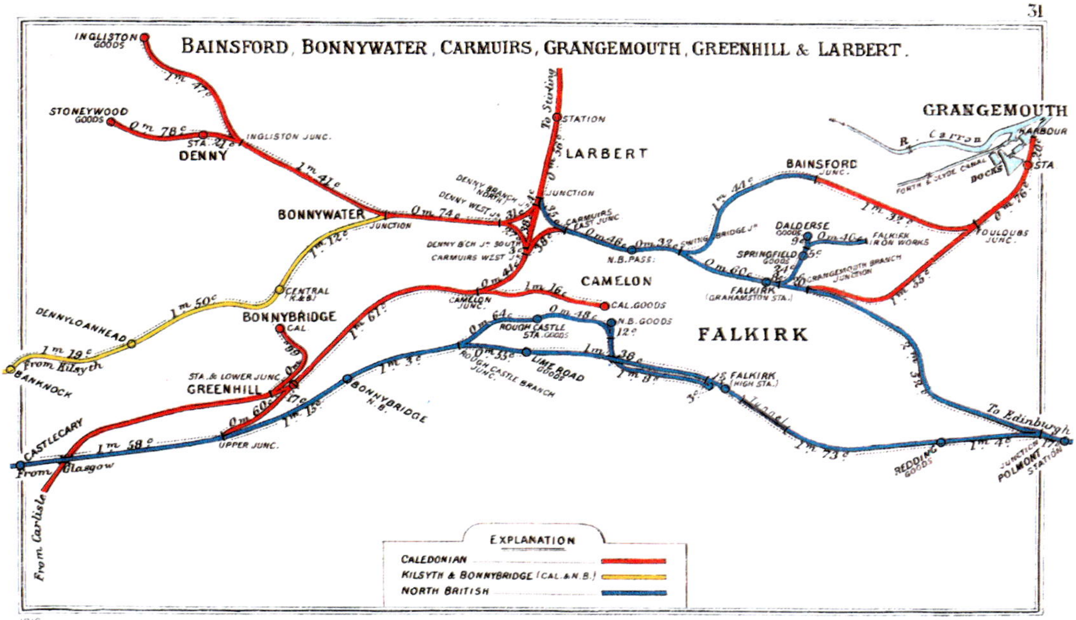

A Clearing House map showing the area around Bonnybridge.

The connection with the Caledonian was made on 7 August 1848 at Lower Greenhill station, seen here in 1917. Lower Greenhill closed on 18 April 1966. There was also an Upper Greenhill station, located where the E&G and SCR lines met. This station was open between 1848 and 1865.

The Caledonian opened a branch to Bonnybridge Canal station in 1886. A shuttle service of three trains a day operated from Greenhill. It closed to passengers on 28 July 1930 but remained open for freight until 1964.

A westbound coal train behind J38 0-6-0 No. 659223 near Bonnybridge High station. (Ben Brooksbank)

Bonnybridge Central.